I See
Animals
Hiding

Dedicated to Jordan and Landon

I See Animals Hiding

JIM ARNOSKY

SCHOLASTIC
HARDCOVER

SCHOLASTIC INC.
New York

I see animals hiding. I see a porcupine high in a tree.

Wild animals are shy and always hiding. It is natural for them to be this way. There are many dangers in the wild.

Even when they are caught unaware out in the open, wild animals try to hide. They stay behind whatever is available — a thin tree trunk or even a single blade of grass. Most of the time they go unnoticed.

The colors of wild animals match the colors of the places where the animals live. Because of this protective coloration, called camouflage, wild animals can hide by simply staying still and blending in.

Woodcocks and other birds, which spend much of their time on the woodland floor, have patterns and colors like those of dry leaves.

I see animals hiding. I see two woodcocks on the leafy ground.

Of all wild animals, deer are the wariest. Even though their colors are camouflaged, they feel safe only where there are good hiding places nearby.

In a summer meadow of tall grasses and small shrubby trees, deer can hide quickly by just lying down.

There are 20
deer on the
snowy hill.
Can you find
them all?

In autumn, deer shed their red-brown summer coats and replace them with warmer, grayer winter coats that better match the gray and brown trunks of leafless trees.

I see animals hiding. I see a whole herd of deer on a winter hill.

Snowshoe hares change from summer brown to winter white. The only way to spot a snowshoe hare in a snowy scene is to look for its shiny black eyes.

Squint your eyes and you will see just how invisible a snowshoe hare on snow can be.

Here are three more animals that are as white as snow. The arctic fox and long-tailed weasel change from winter white to summer brown. The snowy owl stays white year-round.

Besides an owl, there is one other bark imitator on this tree. Can you tell what it is?

The colors and patterns of screech owls blend perfectly with tree bark. These small owls can sleep all day out in the open and not be discovered.

I see animals hiding. I see an owl and a
moth on a limb.

Trout are camouflaged by color and shape to blend with the smooth mossy stones in a stream.

Looking down in a brook, I see a speckled trout swimming amid speckled stones.

I see animals hiding. I see a garter snake slithering through the grass.

Up close a snake in the grass may be easy to see. But as long as the snake keeps a safe distance from its enemies, it can sneak by, looking like just another broken branch on the ground.

Stand back a few steps from this page, and using only your eyes, try to follow the line of the snake from its head to its tail. Can you tell what is snake and what is stick?

A bittern is a wading bird whose brown
streaks and long sticklike legs naturally
blend in with the cattails and reeds that
grow along shorelines.

When a bittern really needs to be invisible,
it points its bill upward and sways
its long neck, like a cattail swaying gently
in a breeze.

And last
but not least:
Animals hide
by staying
inside.

Copyright © 1995 by Jim Arnosky

All right reserved. Published by Scholastic Inc.

SCHOLASTIC HARDCOVER is a registered trademark of Scholastic Inc.

No part of this publication may be reproduced in whole or in part,
or stored in a retrieval system, or transmitted in any form or
by any means, electronic, mechanical, photocopying, recording,
or otherwise, without written permission of the publisher.
For information regarding permission, write to Scholastic Inc.,
555 Broadway, New York, NY 10012.

Library of Congress Cataloging-in-Publication Data

Arnosky, Jim.
I see animals hiding / by Jim Arnosky.   p.   cm.